W9-ASE-421

Early History of the Electron Microscope

DISCARDED
JENKS LRC
GORDON COLLEGE

HISTORY OF TECHNOLOGY MONOGRAPHS

Early History of the Electron Microscope

Second Edition

L. L. Marton

With a preface by

Dennis Gabor, Nobel Laureate

And a postscript by

Charles Susskind

JENKS L.R.C.
GORDON COLLEGE
255 GRAPEVINE RD.
WENHAM, MA 01984-1895

San Francisco Press, Inc.

Box 426800, San Francisco, CA 94142-6800, USA

Copyright © 1968, 1994 by San Francisco Press
Box 426800, San Francisco, CA 94142-6800, USA

QH
212
.E4
M35
1994

Printed in the U.S.A.

This book is based on a special lecture given by the author at the Institute of Electrical and Electronics Engineers 9th Annual Symposium on Electron, Ion, and Laser Beam Technology held in Berkeley, California, in May 1967.

Cover: Author with his second electron microscope in Brussels (1933-34); his first (1932), mounted horizontally, is visible in the background.

Library of Congress Catalog Card No. 94-67380

ISBN No. 0-911302-73-5

TABLE OF CONTENTS

Preface

Prefaces ought to be read after, not before. Otherwise the writer of the Preface can do little more than the chairman who introduces the speaker, starting with his birth and ending with his honors. But the writer of the "post-face" need not introduce Dr. Marton; he has introduced himself in his charming, modest way, and must have given any reader a vivid impression of a dedicated scientific life, well employed. Only one thing remains to be added, and that is not a little thing: his merits.

The modern electron microscope is, I believe, the most wonderful and the most successful instrument of our times. It has realized the dream of Ernst Abbe, that human ingenuity will break through the barrier of resolving power which he was the first to recognize. The electron microscope has first made visible the bacteriophage, which until then existed only as d'Hérelle's hypothesis; more recently it has resolved the millimeter-long (on a germ scale mile-long) skein of DNA which almost fills the body of this micro-micro-organism. It has shown up viruses and unsuspected details in cells and at synapses, and it has created the new science of micro-anatomy. I know of no other instrument which has given a comparable service to science. Why then has the intellectual achievement of creating it found so little recognition? The modern light microscope took 300 years to develop, from Janssen to Abbe and Zernike. The electron microscope took less than 30 years, and all its creators are still alive, with the exception of Bodo von Borries. But who knows their names, outside the small circle of specialists? Even its users in the hospitals and research centers know usually only the name of the firms that make the instruments: RCA, Siemens, MV, JEOL.

The reason is that *a posteriori* the electron microscope appears to be a very obvious invention. One has only to combine the fact than an axially symmetric electric or magnetic field is an electron lens with wave mechanics in order to see the possibility of an electron microscope, together with its basic limitation, which is far beyond that of the optical microscope. That much was indeed at once obvious to the marvelously quick brain of Leo Szilard, who in 1928 suggested to me that I should make an electron microscope. To this suggestion I gave the answer, which, I believe, would have been given by almost all physicists: "What is the use of it? Everything under the electron beam would burn to a cinder!"

The first pioneers believed as I did that if not everything, *almost* everything would burn to a cinder under the electron beam. So Knoll and Ruska started with the few things that would not "burn": wires of platinum and tungsten; and Marton impregnated his first organic preparates with osmium. "Let it burn, but let us look at the cinder." The great discovery dawned on them gradually, in the course of several years. The electron microscope is similar to the light microscope in its basic design, but entirely different in its action, that is to say in the way it forms contrast in the image. Absorption played almost no part. The main source of contrast was *scattering*, and the first to recognize this fact clearly was Bill Marton. He was also the first to open thereby the way to the biological applications of the electron microscope. But even that was not the end of surprises, because in very thin objects, where even scattering was too weak, phase contrast took over. As Otto Scherzer put it: *"Erst war es Vielfachstreuung, dann war es Einmalstreuung, jetzt ist es Keinmalstreuung."* *

Mother Nature was kind to the courageous pioneers of electron microscopy who embarked on this long trek of discovery. They followed (perhaps without knowing it) the magnificent maxim of William the Silent: *"Nul n'est bésoin d'espérer pour entreprendre, ni de réussir pour persévérer."* † In the end they did succeed, and they lived to see their creation become a powerful tool of medicine and of the biological sciences. That they have not succeeded in obtaining the recognition which was their due from their fellow scientists is a matter of less importance for such dedicated scientists as Bill Marton, and the most cooperative of scientists' wives: Claire Marton.

Dennis Gabor

London, January 1968

---

* First it was multiple scattering, then it was single scattering, now it is no (not-even-single) scattering.

† There is no need to hope in order to undertake, nor to succeed in order to persevere.

Introduction

Some time ago I read a mystery story by Josephine Tey. I do not know whether you like mystery stories—I like the good ones. In the one I invoke, the author reexamines an ancient mystery, applies new evidence and some solid reasoning, and comes out with conclusions diametrically opposed to the popularly accepted ones.

I quote this example as an illustration of the fascination of digging for the sources of our information. In the history of science or technology, as in the history of political or other events, the accepted view of the origin of something new is not necessarily the most accurate. There is a challenge in finding out how far back one can trace an idea. In writing down the history of the electron microscope, I had an opportunity to look up some material that was unknown to me and that makes my story more interesting. Without going into the details of all I found, let me start with an outline of the origins of the subject.

The electron microscope has been called an instrument whose development would have been inconceivable without prior development of wave mechanics. Indeed, if we look only at the diffraction-limited aspects of our instrument, this statement is true, perhaps even if other aspects of the mechanism of image formation are considered, such as contrast through scattering in the object, although there earlier theories accounted for part of the effect. Like any statement that is too sweeping, it is no longer true when we look at the electron-ballistic aspects of electron optics, because they go back to simple Newtonian mechanisms and that is where I should like to start my story.

Beginnings of Electron Optics

Electron optics started before the electron was discovered. This statement is less facetious than it appears at first blush. We must remember that electron optics is not limited to the Gaussian optics of axially symmetrical systems. The first observations of electron-optical phenomena go back to times immemorial, when man first admired the northern light, the aurora borealis. It is no surprise, therefore, that the first electron-optical theory deals with the theory of the aurora borealis—and that in itself is a remarkable story!

The first truly quantitative attempt to create a theory of aurora borealis is contained in the series of papers that the Norwegian mathematician Fredrik Carl Mülertz Størmer (1874-1957) began publishing in 1907. He assumed that the earth's magnetic field could be represented by a magnetic dipole and that the electrons coming from outer space would produce ionization in this dipole field.[1] The idea was not entirely new, as another Norwegian, the physicist Kristian Olaf Bernhard Birkeland (1867-1917), had been led 10 years earlier by his experiments on cathode rays to the idea that the aurora was due to charged particles sent out from the sun and interacting with the earth's magnetic field.[2] Størmer in his papers gave credit to Jules Henri Poincaré (1854-1912) for first integrating in 1896 the equations of motion of cathode rays in a magnetic field. Poincaré on his part refers to the mathematical treatment of the equilibrium shape of a weightless and flexible conductor carrying a current and suspended in a magnetic field.[3] This work had been done by Jean Gaston Darboux (1842-1917) in 1878 and provides the mathematical tool for treating the trajectories of cathode rays in a magnetic field.[4]

Let us now look at two ingredients of the theory of the aurora. One is the linking of the aurora to the earth's magnetic field. Without going into details, we may mention some of the illustrious men who contributed to this facet: going backward in time we have Joseph Henry (1797-1878), Karl Friedrich Gauss (1777-1855), Benjamin Franklin (1706-1790), and finally (perhaps) the astronomer Edmund Halley (1656-1742), discoverer of Halley's Comet. As far as I can tell he was the first to observe the corre-

---

1. C. Størmer, "Sur les trajectories des corpuscules électrisés dans l'espace," *Arch. sci. phys. nat. Genève* (4)24: 5-18, 113-158, 221-247, 317-354, 1907; 32: 117-123, 190-219, 277-314, 415-436, 501-509, 1911; 33: 51-69, 113-150, 1912. See also C. Størmer, *The Polar Aurora*, Oxford: Clarendon Press, 1955, pp. 18ff, 209ff.

2. K. Birkeland, "Sur les rayons cathodiques," *Arch. sci. phys. nat. Genève* (4)1: 497-512, 1896.

3. H. Poincaré, "Remarques sur une expérience de M. Birkeland," *Compt. rend.* 123: 530-533, 1896.

4. J. G. Darboux, "Problème de mécanique," *Bull. sci. math.* (2)2: 433-436, 1878.

lation between disturbances of the earth's magnetic field and occurrences of the aurora.

The second, mentioned earlier, is Birkeland's assumption that the charged particles come toward the earth from the sun. Again, without looking at intermediate steps, we find the elements of that idea in an 18th century book by Jean Jacques d'Ortous de Mairan (1678-1771), who attributed the origin of the aurora to "solar matter" entering the earth's atmosphere.[5]

After this demonstration that no discovery or invention exists without quite a bit of prehistory, let us proceed with our more limited subject: the electron-optical experiments and theories leading to the electron microscope.

Although the existence of interesting light phenomena occurring under electrical excitation of an evacuated vessel were known since 1706 (Francis Hauksbee, d. 1713), the first observation of rectilinear propagation of the rays emitted by the cathode had to wait until 1859 (Julius Plücker, 1801-1868). That focusing of the rays can occur was first shown when William Crookes (1832-1919) employed a concave cathode and made use of the fact that the trajectories appeared to be normal to the cathode surface. That was not quite yet the transmission-type optics so prevalent in the development of the electron microscope, but it pointed toward light-optical analogies.

Without recognizing the optical properties, the early users of cathode-ray oscillographs were the first to employ axially symmetrical magnetic or electrostatic fields for the purpose of concentrating their beams. Three names stand out at the turn of the century: Johann Emil Wiechert (1861-1928), Harris Joseph Ryan (1866-1924), and Arthur Rudolf Berthold Wehnelt (1871-1944). According to Gabor's excellent brief history of the electron microscope, even a quarter of a century later, when he invented the first iron-clad magnetic lens, his only purpose was to concentrate the beam in the oscillograph he was building.[6] The strange thing is that even the derivations of Hans Walter Hugo Busch in 1926-27, which are considered as the formal foundation of electron optics and clearly demonstrated the analogy between an axially symmetrical magnetic field and a lens, did not lead immediately to the development of electron-optical experiments.[7] In that connection, Gabor tells an interesting anecdote.

---

5. J. J. d'Ortous de Mairan, *Traité physique et historique de l'aurore boréale*, Paris: Imprimerie Royale, 1733.

6. D. Gabor, *The Electron Microscope*, London: Hulton Press, 1944, pp. 5-6.

7. H. Busch, "Berechnung der Bahn von Kathodenstrahlen im axialsymmetrischen electromagnetischen Felde," *Ann. Phys.* (4)81: 974-993, 1926; and "Über die Wirkungsweise der Konzentrierungsspule bei der Braunschen Röhre," *Arch. Electrotechn.* 18: 583-594, 1927.

When Busch worked out the electron trajectories in axially symmetric fields, he did not think either of wave mechanics, or of the Hamiltonian analogy. Only when he had finished his paper, and showed it to a theoretical physicist, did he hear the exclamation: "What a nice illustration of the Hamiltonian analogy!" It may be remembered that Sir William Hamilton wrote his celebrated Memoirs almost exactly a hundred years before. It is amusing to think that during those hundred years many hundreds of able students studied the Hamiltonian theory of dynamics, and apparently there was not one to ask himself: "Well, if there is such a close analogy between dynamics and optics, what is the dynamical analog of a lens?" Presumably for most students, the Hamiltonian analog was merely something to be acknowledged with a passive nod en route to the esoteric mysteries of Canonical Transformations and the Last Multiplier.[8]

The first conscious applications of electron-optical elements were in components of oscillographs by Brüche[9] and Zworykin.[10] How and when the idea of a microscope using electrons in combination with electromagnetic fields originated is something of a mystery. Again, Gabor tells us he had conversations on this subject as early as 1928.[6] In 1931 several events occurred which may be considered as the true birth of our instrument and which may be therefore discussed at some length.

---

8. D. Gabor, "Die Entwicklungsgeschichte des Elektronenmikroskops," *ETZ* 78: 522-530, 1957.

9. E. Brüche and O. Scherzer, "Die Braunsche Röhre als elektronoptisches Problem," *Z. techn. Phys.* 14: 464-466, 1933; and E. Brüche, in W. Petersen, Ed., *Forschung und Technik*, Berlin: Springer, 1930.

10. V. K. Zworykin, "Electron optics," *J. Franklin Inst.* 215: 535-555, 1933.

First Electron Microscopes

Two dates, very close together, are significant. On 4 June 1931 Max Knoll gave a colloquium talk at the Technische Hochschule in Berlin, in which he reported on experiments undertaken jointly with his student Ernst Ruska. Using experiences gained with magnetic lenses in the years 1928-1930, they built a two-lens microscope (Fig. 1), which was demonstrated to a number of visiting scientists between February and May of that year. They obtained magnified images of simple objects, such as a T-shaped aperture and a coarse wire mesh.

On 30 May 1931 of the same year the research director of the German company Siemens-Schuckert, Günther Reinhold Rüdenberg (1883-1961), filed a patent application with the German Patent Office on the subject of combining several electron lenses, magnetic or electrostatic, for the purpose of using the combination as an electron microscope.[11]

This indeed is a strange coincidence! There is no doubt about preliminary work having been done at the TH Berlin under Knoll's direction for several years, because that is documented by the existing thesis work of Ruska in 1929 and 1930, in addition to the well-documented and extensive paper by Knoll and Ruska, which was submitted to *Annalen der Physik* on 10 September 1931.[12]

A year after filing for his patent, on 7 June 1932, Rüdenberg wrote a letter to the editor of *Naturwissenschaften,* reproduced here in full:

Since various proposals for the construction of electron microscopes have lately appeared from various sources, I wish to point out that work in this direction has also been under way for some time at Siemens, to use magnetic or electric fields in microscopes or telescopes for electron or proton beams. Our goal is above all to produce greatly magnified images of submicroscopic objects at rest or in motion. If they may not be exposed to high vacuum, they can be illuminated and observed through transparent windows and the display on the luminous screen can be further magnified by an optical microscope as needed. Although our basic patents go back to May 1931, we do not intend to present detailed publications until practical development has been carried further.[13]

No such publication was ever issued, to my knowledge.

---

11. R. Rüdenberg, German patent application, 30 May 1931; British patent 402 781; U.S. patent 2 058 914; and French patent 737-716, which was actually the first to be issued (on 15 December 1932).

12. M. Knoll and E. Ruska, "Beitrag zur geometrischen Elektronenoptik," *Ann. Phys.* (5)12: 607-661, 1932.

13. R. Rüdenberg, "Elektronenmikroskop," *Naturwiss.* 20: 522, 1932. The articles to which he refers are M. Knoll and E. Ruska, *Ann. Phys.* (5)12: 607 and 641, 1932; E. Brüche, *Naturwiss.* 20: 49, 1932; F. Hamacher, *Arch. Elektrotechn.* 24: 215, 1932; and H. Johannson, *Naturwiss.* 20: 353, 1932.

FIG. 1.—First two-stage electron microscope of Knoll and Ruska. (Taken from a paper published in 1932.[12])

The same year saw two other efforts. One is a short publication by Brüche in Naturwissenschaften, written in November 1931 and published the following January, reporting on the development of an electrostatic emission microscope. He says that the research laboratories of the Allgemeine Elektrizitäts-Gesellschaft (AEG) in Berlin also "have been engaged in work in the same direction [as Knoll and Ruska] for about a year."[14]

And in far-off California, the American Physical Society held a meeting in Pasadena on 15-20 June 1931. Two physicists from the Bell Telephone Laboratories, Davisson and Calbick, presented a paper on electron lenses, an abstract of which appeared that summer.[15] They gave the focal length of a circular hole in a flat plate, perhaps the basic electron-optical configuration—the electrostatic aperture lens. (The importance of this contribution is not lessened by the fact that the formula was in error by a factor of 2, as the authors themselves presently pointed out.[16])

Another interesting point is a semantical one. In an earlier paper by Ruska and Knoll, submitted for publication on 28 April 1931, the authors do not mention electron optics, electron lenses, or electron microscopes.[17] Neither does the Rüdenberg patent application, which was filed on May 30. However, in June Davisson and Calbick use the words "electron lenses" and the paper submitted by Ruska and Knoll on September 10 is entitled "Contributions to geometrical electron optics."[18] It is not clear who invented our present terminology and when, but it must have happened on both sides of the Atlantic practically simultaneously during the summer of 1931.

One clue may be supplied. In September 1928, Lester Halbert Germer, Davisson's collaborator at the Bell Telephone Laboratories, had written a paper entitled "Optical experiments with electrons." In it he said:

> The title of this paper may sound paradoxical. Almost everyone knows that electrons are discrete units of electricity; they are the indivisible particles of which atoms are made. How then can optical experiments be performed with them?
> In spite of all it must be emphasized that the experiments with electrons which are to be described here are *really optical experiments*. In no sense are the phenomena which Dr. Davisson and I have observed to be thought of as only vaguely resembling optical phenomena. The experiments which we have performed are actually optical experiments with electrons, just as certainly as experiments with x-rays are optical experiments. The technic of x-ray experimentation

---

14. E. Brüche, "Elektronenmikroskop," *Naturwiss.* 20: 49, 1932.

15. C. J. Davisson and C. J. Calbick, "Electron lenses," *Phys. Rev.* 38: 585, 1931. Clinton Joseph Davisson (1881-1958) was co-winner (with George Paget Thomson) of the Nobel Prize in physics in 1937, "for the experimental discovery of the interference phenomenon in crystals irradiated by electrons."

16. C. J. Davisson and C. J. Calbick, "Electron lenses," *Phys. Rev.* 42: 580, 1932.

17. E. Ruska and M. Knoll, "Die magnetische Sammelspule für schnelle Elektronenstrahlen," *Z. techn. Phys.* 12: 389-399 and 448, 1931.

18. Knoll and Ruska, *op. cit.*

is, of course, quite different from the technic of experimentation with visible light. The technic of our experiments with electrons is also quite different from the technic of ordinary optics. This difference is, however, a difference in experimental convenience rather than a difference in principle. Our experiments establish the wave nature of moving electrons with the same certainty as the wave nature of x-rays has been established.[19]

19. L. H. Germer, "Optical experiments with electrons," *J. Chem. Ed.* 5: 1041-1055 and 1255-1271, 1928.

I Become Interested

When I became involved in the subject, the status of electron optics and electron microscopy could be summarized as follows. It was known that far-reaching analogies existed between the behavior of light and of electrons in their respective refractive media, as well as with respect to diffraction phenomena. The geometrical optics of electrons had been explored with regard to the construction and properties of magnetic and electrostatic lenses and primitive compound microscopes had been built and tested. The elementary theory of both kinds of lenses had been developed and some thought had been given to the optical aberrations of such lenses. The experimental work was greatly influenced by oscillograph development: the work of Knoll and Ruska at TH Berlin was a direct continuation of their studies of the high-voltage cathode-ray oscillograph, whereas Brüche and his collaborators at AEG were more interested in the low-voltage sealed-tube devices. The competition between the two groups had very beneficial effects on the development of geometrical electron optics.

At that time I was a member of the faculty of the Université Libre in Brussels. Following my thesis work on infrared spectroscopy, I got involved for some time in photoelectricity, did a little work on x rays, and came back again to infrared. By the summer of 1932 I had also read most of the relevant literature on electron optics. Not all: I had missed one or two papers, and the Rüdenberg patent application was not to be available until some time later. I got quite excited and as I was still interested in photoelectricity, it occurred to me immediately that one might establish a connection between these two branches of physics. It so happened that an interesting piece of work came to my attention at that time implying the possibility of a patch effect in photoelectric emission, with adjacent areas having somewhat different work functions. My first reaction was: if there is anything in that proposal, we should be able to see the results by means of this splendid new tool, the electron microscope.

I was fortunate enough to have a department chairman who coupled a broad outlook with a fine sense of experimental possibilities. Professor Émile Henriot (1885-1961) became a good friend and I could freely talk over my ideas and my problems with him. He immediately encouraged me and, although the laboratory resources were rather meager, gave me the means for a first try. The result was my first primitive electron microscope, which was in operation by the beginning of December 1932.

The configuration was quite simple. The main body was a brass tube about 2 in. in diameter and about 1 ft in length. The single lens consisted of a wirewound coil that could be slid along the brass tube. The electron source was a flat tungsten coil, with the brass tube serving as an anode. At the opposite end a fluorescent screen was used either for visual observation or for external photography. Inside the brass tube a specimen mount was placed with means for positional changes. Needless to say, for changing specimens or for changing their position, the instrument had to be opened to the atmosphere. (At the time of writing, a replica of this first primitive microscope is being erected as a working exhibit at the Museum of History and Technology, Smithsonian Institution, Washington.)

Primitive as it all was, the first few pictures showed reasonably good imaging quality (Fig. 2), which encouraged me to undertake the construction of a better instrument. Although I never achieved my original purpose of doing photoelectric work, in the course of making adjustments on my instrument I observed images of the tungsten spiral, which excited my curiosity. As seen in Figure 3, the emission along the various parts of the spirals is not uniform. Since I had done some work a few years earlier on tungsten and was familiar with its crystalline appearance, my immediate interpretation was that different crystalline faces present at the surface emit with different work functions and that the difference in emission is due to differences of this kind.

About the same time investigations started at the AEG laboratories on the emission of tungsten and I refrained from publishing my first results, although they were at least as convincing as those from Berlin. The difference was that an industrial laboratory had much ampler means for supporting and advancing research of this sort than a small Belgian university. A little more about this subject later.

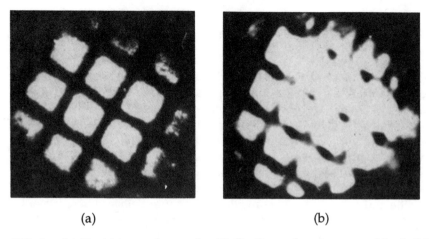

(a)   (b)

FIG. 2.—(a) First image observed with the Brussels microscope No. 1; (b) improved image observed with the same microscope.

FIG. 3.—Electron micrograph of emission from a tungsten spiral (1933).

Progress in Germany

The AEG laboratories were not the only ones doing emission microscopy. At the TH Berlin, Knoll and his collaborators, Fritz Georg Houtermans (1903-1966) and E. Schulze, undertook the investigation of oxide cathodes with a two-stage magnetic instrument.[20] Both the electrostatic instrument of Brüche and Johannson[13] (single stage) and the magnetic microscope of Knoll et al. operated with external photography and at low magnifications (below 100x). Both had a horizontal structure, with optical-bench components. However, the optical-bench arrangements of Brüche and Johannson were inside the vacuum belljar, whereas Knoll, Houtermans, and Schulze arranged them outside the vacuum around an elongated glass tube that served as the microscope housing.

In the same year, 1932, Knoll and Ruska published a description of their first two-stage vertical microscope.[21] It was a considerably more advanced instrument, consisting of a gaseous-discharge (cold-cathode) electron gun, a condensor lens, and objective and projection lenses. The fluorescent screen could be observed through a mirror mounted behind a viewing port, which caused some distortion in photographing the final image from the outside. The operating potential difference on the gun appears to be 65 kV, although the paper says that the instrument was built for a range of 10 to 100 kV. The highest magnification illustrated was 150x, without any indication of resolution. However, the paper contained the first discussion of resolution limitations based on the Abbe relation: the estimated numerical aperture of 0.02 led to the somewhat optimistic (theoretical) limits of 15 Å for 1.5keV and 2.2 Å for 75keV electrons. Knoll and Ruska also discussed some of the possible perturbations, such as space charge, scattering from gas molecules when operating at too high pressures, external ac or dc fields, fluctuations of the high voltage, and mechanical perturbations such as vibrations. Although the discussion was qualitative, it performed the useful purpose of calling attention to these problems. A short section of the paper was devoted to an abbreviated discussion of the possible advantages of an ion microscope.

During this time the group at AEG laboratories undertook a detailed investigation of oxide cathodes. Brüche and Johannson used a time-compression technique for a motion-picture presentation of the activation process. They used their original single-stage electrostatic microscope

20. M. Knoll, F. G. Houtermans, and W. Schulze, "Untersuchung der Emissionsverteilung in Glühkathoden mit dem magnetischen Elektronenmikroskop," *Z. Phys.* 78: 340-362, 1932.

21. M. Knoll and E. Ruska, "Das Elektronenmikroskop," *Z. Phys.* 78: 318-339, 1932.

and replaced the still camera in front of the fluorescent screen by a movie camera. Individual exposure times were 3 to 6 s; a 2hr activation process was compressed into 2 min when projected. A little later the investigation was extended into the observation of pure-tungsten and thoriated-tungsten cathodes, with special attention to the activation of the latter.[22] AEG was, after all, in the vacuum-tube business and improvements in cathode emission were of immediate economic interest.

At the end of the latter paper there is a short note by Johannson reporting on the first calculation of the focal length of an "einzel" lens (a three-electrode configuration in which the two outer electrodes are at the same potential), done jointly with Otto Scherzer. This is the first appearance of Scherzer's name in the electron-optical literature. Later, together with Walter Glaser (1906-1960), he became one of the outstanding theoreticians in the field.

Glaser's first theoretical papers in electron optics appeared in 1933. He had a classical education in optics and was quick in applying the eiconal concept to this new problem. He showed in his very first paper that the formalism derived from the Fermat-Hamilton analogy is useful in describing the imaging properties of electron-optical systems[23] and proceeded in one of his next papers to apply that result to a calculation of image aberrations in an electron microscope, showing the increased importance of the spherical aberration (as compared to light optics) and also predicting added aberrations not known to light optics.[24]

While Glaser was busy exploiting the approach derived from Seidel's classical theory, Scherzer developed the trajectory method of electron-optical theory.[25] Both methods have proved to be very fertile, the first by its mathematical elegance, the second by its adaptability to physical-model representation. A book entitled *Contributions to Electron Optics*, edited by Busch and Brüche, appeared in 1937. The two summaries by Glaser and by Scherzer, which are abbreviated descriptions of the two methods, remain in my opinion even today the most concise presentations of the outlines of the two approaches to electron-optical theory.[26]

---

22. E. Brüche and H. Johannson, "Kinematographische Elektronenmikroskopie von Oxydkathoden," *Ann. Phys.* 15: 145-166, 1932.

23. W. Glaser, "Über geometrisch-optische Abbildung durch Elektronenstrahlen," *Z. Phys.* 80: 451-464, 1933.

24. W. Glaser, "Zur geometrischen Elektronenoptik des axialsymmetrischen elektromagnetischen Feldes," *Z. Phys.* 81: 647-686, 1933; and 83: 104-122, 1933.

25. O. Scherzer, "Zur Theorie elektronenoptischer Linsenfehler," *Z. Phys.* 80: 193-202, 1933.

26. W. Glaser, "Elektronenbewegung als optisches Problem," in *Beiträge zur Elektronenoptik* (H. Busch and E. Brüche, Eds.), Leipzig: Barth, 1937, Chap. 5, pp. 24-33; and O. Scherzer, "Berechnung der Bildfehler dritter Ordnung nach der Bahnmethode," *ibid.*, Chap. 6, pp. 33-41.

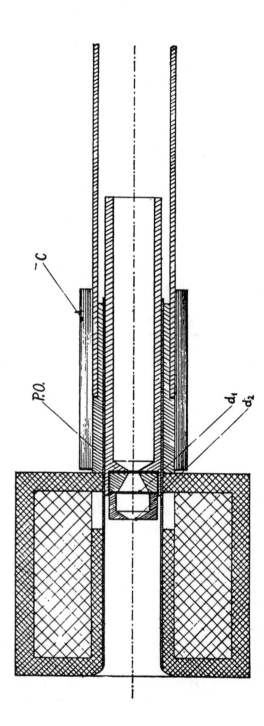

FIG. 4.—Objective and specimen holder of Brussels microscope No. 2.

I can now come back briefly to my earlier remark that wave mechanics had to exist before the development of the electron microscope was possible, and amplify that statement. Although Glaser's approach starts with the assumption of the wave nature of the electron, he found it necessary 20 years later to develop a wave-mechanical approach to electron-optical theory *ab initio*. This need was a consequence of the fact that the whole treatment after the initial assumption was completely classical, in fact just as classical as the trajectory method of Scherzer.

In the meantime the number of papers on the subject of electron microscopy started increasing and I cannot possibly comment on all of them within the framework of this review. I shall concentrate instead on papers that were more closely related to my own work, which accelerated after a slow start. The year 1933 was spent in exploring the limitations of my first instrument, constructing the second one, publishing one short review paper on electron optics (jointly with Maurice Nuyens[27]), and most important, acquiring a lifelong collaborator in the person of my wife, Claire.

My new instrument[28] had a cold-cathode (gaseous-discharge) electron gun, condensor, objective and projector lenses, a translucent metallized fluorescent screen, and for ease of photography of the screen—horizontal arrangement of the components. It is interesting to note that the projector lens was mounted with a sliding fit on the *outside* of the brass housing of the microscope, with exchangeable polepieces *inside* the vacuum. The new instrument of Ruska, whose description appeared early in 1934[29] (when my second instrument was undergoing its first tests), used separate polepieces in a somewhat similar manner (Fig. 4). A rubber sleeve, fitting tightly over the brass tubing connecting the gun with the microscope proper, allowed a relatively easy change of specimens. In view of the beginning race toward improved resolving power I set myself the goal of concentrating on two tasks: first, to explore some of the possible applications of the instrument and to make it the most useful for such purposes; and second, to try to understand the mechanism of image formation.

---

27. L. Marton and M. Nuyens, "Meetkundige optica der elektronen," *Wis. en natuurkundig tijdschr.* 6: 159-170, 1933.

28. L. Marton, "La microscopie électronique des objets biologiques," *Bull. Acad. Roy. Belg.* 20: 439-446, 1934.

29. E. Ruska, "Über Fortschritte im Bau und in der Leistung des magnetischen Elektronenmikroskops," *Z. Phys.* 87: 580-602, 1934.

First Applications to Biology and Image Interpretation

I came very quickly to the conclusion that biological objects would offer the richest fields for this new tool. At the beginning I was overly impressed by the arguments of the German school, according to which the heating of the specimen by the energy lost by the electron beam in passing through would "burn it to a cinder." I began by seeking a means for heat-resistant impregnation of the biological object that would serve as a kind of replica of the supposedly heat-sensitive material. Another possibility was that such impregnation might render the object less destructible. A third alternative was intense cooling by means of a support with a high heat conductivity; or one might use a combination of any of the above methods. Fortunately some of my friends at the university were good biologists and the botanist Marcel Homès (later president of the Royal Academy of Belgium) and the zoologist Paul Brien gave me some microtome slices fixed with osmium tetroxide. In view of the then available microtome technique, only relatively thick cuts were available and the expected energy losses were high. Nevertheless, the technique was sufficiently successful to produce the first histological photographs ever produced by the electron microscope (Fig. 5). The date of these first micrographs was 4 April 1934; they were presented at the 8 May 1934 meeting of the Belgian Academy[28] and first appeared in print in the 16 June issue of *Nature*.[30]

Soon thereafter I tried the alternative of cooling by placing the histological preparation on an aluminum foil 0.5 µm thick. Owing to the added scattering there was considerable loss of contrast, but a gain in internal details. In the first process only cell walls had remained coherent enough to produce an image, but now we could see cell nuclei too.[31]

Simultaneously with these results I slowly attacked the problem of interpretation of the image. I felt that there was little advantage in higher resolution if we did not know what the image represented. Available in the literature were two statements. In 1933 von Borries and Ruska had written that ". . . metallic foils . . . permit by their small density and thickness changes (different strong absorption) the recognition of a 'structure'."[32] Eight months later Ruska wrote: "The . . . micrographs show the greatly improved optical performance of the new apparatus, without an

---

30. L. Marton, "Electron microscopy of biological objects," *Nature* 133: 911, 1934.

31. L. Marton, "Electron microscopy of biological objects," *Phys. Rev.* 46: 527-528, 1934.

32. B. v. Borries and E. Ruska, "Die Abbildung durchstrahlter Folien im Elektronenmikroskop," *Z. Phys.* 83: 188, 1933.

FIG. 5.—First biological electron micrograph.

attempt to solve the problem of interpretation of the image." In the same paper, however, he says earlier: "One could speak . . . of absorption images and of diffusion images, although diffusion phenomena predominate . . ."[29] If he had written in French, where *diffusion* means both diffusion and scattering, this would be a relatively modern statement and it has been interpreted as such by some writers. In German, however, the word for scattering is *Streuung*. As he used the German word *Diffusion* I am inclined to believe that the first clear statement ascribing the contrast in the image to differences in the scattering properties of the object was contained in my paper of 8 May 1934, where I calculated and tabulated mean scattering angles as a function of thickness and of electron velocity for two substances, beryllium and aluminum.[28] The calculation assumed multiple scattering and showed the need for very thin objects or for electrons of higher energy. It was based on Bothe's formulation of multiple Rutherford scattering for small angles.[33]

---

33. W. Bothe, "Durchgang von Elektronen durch Materie," in *Handbuch der Physik*, Berlin: J. Springer, 1933, 2nd ed., vol. 22, pt. 2, pp. 1-74.

Evolution of the Compound Instrument

Having referred earlier to Ruska's 1934 paper[27] I must say a little more about it. It was submitted on 12 December 1933 and appeared early in 1934. It described the construction and some of the results with a new instrument with improved components. Like all of Ruska's instruments, it was a vertical one (Fig. 6). The gun was very similar to earlier ones. The object chamber was provided with eight positions for that many objects, which could be brought in front of the objective lens through external controls. The construction details of the specimen mounts were carefully worked out to assure optimum heat conductivity. One of the reasons for the multiple object chamber was to provide means for focusing on dummy objects, thus reducing the heat load on the real object. The estimate resolution of this instrument is given as 500 Å. The numerical aperture is calculated to be 0.0001, as contrasted with the geometrically expected 0.08.

Ruska's second paper is devoted to the description of the magnetic objective lens.[34] It had two windings and two gaps. A set of polepieces could be attached in front of one of the gaps and the other gap was short-circuited. For greater constancy the lens was water cooled, a design that was adopted in practically all lenses constructed by Ruska. The whole design and construction shows excellent engineering, a practice that characterizes all Ruska, von Borries, and Siemens instruments.

Very careful measurements were carried out on this lens and the results were compared with those obtained on a bare coil (without iron enclosure), as well as with the theoretically expected optical behavior. One of the important observations was the existence of a minimum focal distance, as well as the behavior of the spherical aberration as a function of the inside diameter of the polepieces.

Toward the end of June of 1934 my wife and I visited Berlin and became personally acquainted with Knoll, Ruska, Brüche, and several others active in electron optics. Ruska was getting ready to take an industrial job and with his going, electron microscopy at the TH Berlin came to an almost complete standstill. When I showed my micrographs of biological objects, it was very gratifying to see initial unbelief ("This is not possible!") give way gradually to requests for prints.

I have looked up the dates of this trip to Berlin because of some of the circumstances surrounding it. The atmosphere in Germany was quite electric. One of our friends (not connected with electron microscopy) took us miles out of Berlin in his car to a place where one could see at least a

---

34. E. Ruska, "Über ein magnetisches Objektiv für das Elektronenmikroskop," *Z. Phys.* 89: 90-128, 1934.

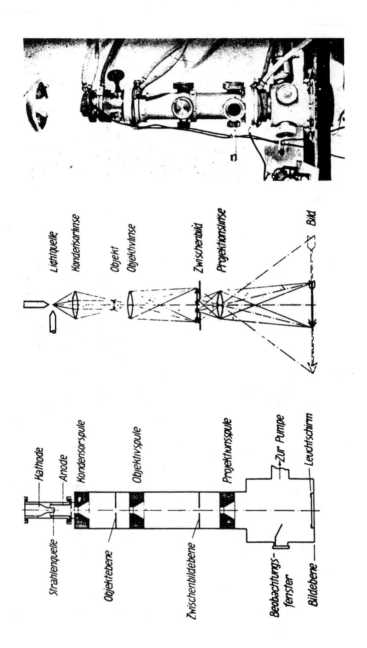

FIG. 6.—Ruska's 1934 microscope.

mile in each direction, stopped there, closed all the windows, and said: "*Now* we can talk." Small wonder, the date was 30 June 1934, the "Night of the long knives"—the great bloodbath of Nazi Germany, when Himmler's *SS* massacred the leadership of Roehm's rival *SA*.

The year 1934 did not end without an important landmark: the appearance of the first major publication in electron optics. Brüche and Scherzer produced a book entitled *Geometrische Elektronenoptik* (Geometric Electron Optics), which remained for many years the standard text for this branch of physics.[35]

The experience gained with my second microscope pointed toward a need for improvement both in technique and instrumentation. It was obvious from the beginning that a reduction of heating of the specimens by the beam used for the observation had a very high priority. One way was to reduce the thickness of the specimens. As to histological objects, the creation of better microtomes was still in the future. In the absence of good microtomes, we had to look for reduction of the radiation load required for focusing and recording the image. As long as we were confined to external photography of the fluorescent screen, the only ways were to improve the quality of the phosphors used, to match the color sensitivity of the photographic material to the light output of the phosphor, and to use more efficient photographic optics. All I could gain by these combined methods was a factor of 50, which was perhaps impressive but still not sufficient. Clearly, we had to go to internal recording of the electron micrographs if we wanted to make progress.

One of the annoying and time-consuming limitations of the first microscopes was the need to open the instrument to air for insertion of new specimens. Although the instruments operated at what we would consider nowadays a relatively high pressure (owing to the use of gaseous discharges in the electron guns), the pump-down time with the inefficient pumps of the day led to a serious interruption of operation for specimen change.

In thinking about internal recording of the image, the same problem arose. If we had to evacuate the whole microscope for each plate insertion, the down-time of the instrument would be excessively long. Obviously airlocks for both specimens and photographic material were the answer. These were the times before rubber gaskets were extensively introduced into vacuum practice. Both airlocks of my new instrument used greased tapered joints, although at one place in the specimen airlock I used one gasket even then. The conception of the specimen airlock allowed a motion of the specimen in two orthogonal directions for the

35. E. Brüche and O. Scherzer, *Geometrische Elektronenoptik*, Berlin: J. Springer, 1934.

first time, so that a crude mechanical stage was available. In order to reduce the pump-down times further, the airlock provided for the insertion of six specimens, which could be exchanged without opening of the airlock. External scales on all motions of the "stage" allowed the reproduction of any stage position within the limited accuracy of the device.

The photographic airlock replaced exposed plates one by one, but permitted the movement of the plate within the exposure chamber for several partial exposures; the fluorescent screen served both as a shutter and as a shading device. The volume of the fore-pumping chamber was very small (only a few cubic centimeters) to reduce the time required for replacing the plates.

Another innovation was an electronic shutter. It consisted of an adaptation of the deflection plates of cathode-ray oscillographs for blocking the beam when not wanted and allowing short time exposures by means of a thyratron-operated timer. The idea was, as mentioned before, to focus on a dummy object, deflect the beam, move the real object into position, and irradiate it only for the minimum time needed to record the image.

Another variant of the same idea was the compilation of a nomogram for presetting all the lenses of the instrument. The results were disappointing. The main obstacle was the presence of grease and other insulating surfaces in and near the beam path, which charged up. This uncontrollable charging not only caused beam deflections but also produced instabilities of the image. By leaving the beam on long enough, we could establish equilibrium conditions for a sufficiently long time to record a reasonably stable image. If the beam was switched on and off, it took several minutes each time to reduce the shifting of the image to acceptable limits.

The construction of the third microscope was started early in the autumn of 1934. By the beginning of winter I could report some progress,[36] although each bit of progress cost a tremendous amount of effort. University funds were practically nonexistent and with the meager resources I had (thanks to family support), Mrs. Marton and myself became frequent visitors at the flea market. A reasonable amount of German equipment was available that had been abandoned after the war of 1914-1918 and numerous components of the new microscope benefitted from this availability at very low prices. The only other source of money was a modest grant from the Institut International de Physique Solvay. Professor Henriot was at that time permanent scientific secretary of the Solvay Congresses in Physics and thanks to his intervention I was for some years the recipient of a Solvay fellowship.

---

36. L. Marton, "Le microscope électronique et ses applications," *Bull. Soc. Franç. Phys.* No. 364, 21 December 1934.

Attempts to Improve Biological and Other Techniques

At about that time I had an opportunity to report on the electron microscope, on the results achieved until that time, and on possible expectations, to an audience composed mostly of medical and biological scientists.[37] After I presented what I believed to be a good case for the new tool, the eminent bacteriologist and Nobel Prizewinner Jules Bordet (1870-1961) rose and said: "Oh no, no! Let us not have an electron microscope—it's troublesome enough to interpret the images we get by the light microscope!"

As the biological specimens required some supporting structure, an important aspect of my work was to improve the nature of such supports. Metallic films did not seem to be the right answer, evaporation techniques were in their infancy, and the only available films were beaten gold foil or aluminum foil. Their unevenness, coupled with their thickness, made them undesirable, but fortunately I remembered a technique developed for infrared windows: the casting of collodion foils onto water surfaces. That gives reasonably good films at the desirable thickness range and to my great pleasure they turned out to stand up very well in the electron beam, thus proving that the heavy-metal impregnation of our specimens was not absolutely necessary, provided the thickness of the layer was sufficiently reduced. Encouraged by this success, my good friend Serge Herzen and I explored a great number of other substances and solvents, but without new results.

Another try in a different direction was less fruitful. I wondered if it would be possible to examine living specimens. The most important limitation appeared to be the dehydration of the specimen in the vacuum of the microscope. Therefore, I tried a separation of the gun part of the instrument from the optical part and inserted Lenard windows made of aluminum foil at the two ends. A living specimen was placed between the two Lenard windows set at a small distance. Needless to say, I could see nothing; the scattering produced by the windows and by the air between them reduced the contrast to zero. Nevertheless, this was the first attempt, which was repeated later, time and again, and which should lead sometime in the not too distant future to interesting results. In particular, I would expect such *in vivo* observation to give new information on the phenomena of cell division and on the behavior of chromosomes during the mitotic process.[38]

---

37. L. Marton, "Le microscope électronique," *Ann. et Bull. Soc. Roy. Sci. Med. et Nat. de Bruxelles* (Nos. 5-6): 92-106, 1934.

38. L. Marton, "La microscopie électronique des objets biologiques: II," *Bull. Acad. Roy. Belg.* 21: 553-564, 1935.

To summarize my progress in 1934, the first biological electron micrographs were produced; the rudiments of biological technique were developed; the technique of thin-film support for any specimen was established and the first proof was found that organic materials stand up in the electron beam, provided they are sufficiently thin; airlocks for specimens and photographic material were created; and last but not least, the beginnings of a quantitative approach to the mechanism of image formation were sketched.

Meanwhile the German work progressed too. I shall list a few of the highlights. Von Borries and Knoll made a quantitative investigation of the sensitivities of various photographic emulsions to electron irradiation at varying energies.[39] Herzog in Vienna published an important theoretical paper on the electron optics of two-dimensional fields.[40] Brüche and Knecht investigated the resolution of an emission microscope and came to the conclusion that the poor resolution must be due to contact potentials on grain boundaries of crystals.[41] It took three more years before the true limitation was recognized; a little more about that later.

The emission microscope was applied first by Brüche and Knecht,[42] a little later by Burgers and Ploos van Amstel (of the Philips Research Laboratories), to a study of the $\alpha \leftrightarrow \gamma$ transition in iron.[43] A cinematographic presentation of these phase transformations was prepared by the latter. I still remember the movie as quite interesting.

As a consequence of a talk before the French Physical Society I was asked to prepare a review of the subject. My friend Henriot added a review of the theory of electron optics and the two contributions appeared together in a pamphlet in 1935.[44]

---

39. B. v. Borries and M. Knoll, "Die Schwärzung photographischer Schichten durch Elektronen und elektronenerregte Fluoreszenz," *Physik. Z.* 35: 279-289, 1934.

40. R. Herzog, "Ionen- und elektronenoptische Zylinderlinsen und Prismen: I," *Z. Phys.* 89: 447-473, 1934.

41. E. Brüche and W. Knecht, "Bemerkung über die Erreichung hoher Auflösungen mit dem elektronenoptischen Immersionsobjektiv," *Z. Phys.* 92: 462-466, 1934.

42. E. Brüche and W. Knecht, "Die elektronenoptische Beobachtung von Umwandlungen des Eisens bei Temperaturen zwischen 500 und 1000°C," *Z. techn. Phys.* 15: 461-463, 1934; and "Die elektronenoptische Beobachtung der Eisenumwandlung vom α- in den γ-Zustand," *ibid.* 16: 95-98, 1938.

43. W. G. Burgers and J. J. A. Ploos van Amstel, "Cinematographic record of the $\alpha \leftrightarrow \gamma$ iron transformation as seen by the electron-microscope," *Nature* 136: 721, 1935; and "Electronoptical observation of metal surfaces," *Physica* 4: 5-23, 1937.

44. L. Marton, *Le microscope électronique et ses applications* (together with É. Henriot, *Optique électronique des systèmes centrés*), Paris: Éditions de la Revue d'Optique, 1935.

This talk before the French Physical Society reminds me of an amusing incident that occurred after the talk. My wife and I stayed with friends in a Paris suburb; Pierre Auger and Frederic Joliot lived in the same area. We started walking toward the subway, but on the way entered a bistro where we started playing some of the "fruit machines," or one-arm bandits as they are known in Nevada. To cut a long story short, all four of us lost every sou we had and were obliged to walk all the way back to the suburb after midnight!

The same year, interest awakened in Britain. Martin published a review, which was a forerunner of his own effort to construct an electron microscope.[45]

My new instrument—No. 3—was finally in operation in July 1935.[46] Besides numerous histological specimens I tried to observe a bacteriological one in August 1935. This first try was not successful and it took more than a year and a half before my technique improved sufficiently to get back to bacteria. In the meantime, however, observation of other biological materials improved and I could show by light-microscopical comparisons that the specimen did not suffer, and that the resolution of biological micrographs exceeded that of the light microscope by a factor of 10. Incidentally, the exposure time on most micrographs was of the order of 1/20 to 1/60 s.

45. L. C. Martin, "Electron optics," *Sci. Progress* 115: 426-437, 1935.
46. L. Marton, "La microscopie électronique des objets biologiques: III," *Bull. Acad. Roy. Belg.* 21: 606-617, 1935.

More on Image Formation

At this time I started the calculation of contrast phenomena in the image.[47] They were still based on multiple scattering, as in my earlier paper. The difference was that whereas earlier I had merely indicated the most probable scattering angles for two substances, this time I set out deliberately not to leave the inferences to the imagination of the reader. I assumed an object having either thickness or density variations, an objective aperture providing a limited solid angle, a scattering distribution depending on the constitution of the object, and contrast production by scattering inside or outside of the solid angle delimited by the aperture. As variables I took beam energy, thickness, density (the atomic number and atomic weight), and the solid angle subtended by the aperture. By assuming a photometric minimum in the photographic record, I defined a "depth resolving power" by analogy with the conventional optical resolving power.

Another conclusion of this paper was to recommend "staining" of specimens with substances of higher atomic numbers, where the contrast of the specimen proper was not sufficient. I also showed that contrast decreases with increasing numerical aperture, a conclusion that places a limitation on attempts to increase the resolving power by reduction of the spherical aberration.

The demonstration that the electron microscope can be used for biological purposes encouraged people in Germany to use the instrument constructed by Ruska. A few papers were published and resolving power considerably in excess of that of the light microscope was demonstrated.[48]

Two rather important papers came out during 1936. One was published by Boersch, who has since become one of the most original thinkers in electron microscopy.[49] His important paper called attention to the close relationship between diffraction and image formation and demonstrated for the first time what became known a few years later as selected-area diffraction.

---

47. L. Marton, "Quelques considérations concernant le pouvoir séparateur en microscopie électronique," *Physica* 3: 959-967, 1936.

48. E. Driest and H. O. Müller, "Elektronenmikroskopische Aufnahmen (Elektronenmikrogramme) von Chitinobjekten," *Z. wiss. Mikr.* 52: 53-57, 1935; F. Krause, "Elektronenoptische Aufnahmen von Diatomen mit dem magnetischen Elektronenmikroskop," *Z. Phys.* 102: 417-422, 1936.

49. H. Boersch, "Über das primäre und sekundäre Bild im Elektronenmikroskop," *Ann. Phys.* (5)26: 631-644, 1936 and 27: 75-80, 1936.

The other important paper was by Scherzer.[50] In a theoretical investigation of image aberrations in electron lenses, he arrived among other conclusions at the result that spherical aberration was unavoidable in a conventional electron-optical system. No combinations of electric or magnetic fields would help, because the final formulation of any combined fields always gave a sum of squares, meaning that that sum was always positive. According to his proof, in an axially symmetrical system, in the case of nearly paraxial image formation, all we can hope for is a reduction of the aberration, but not its elimination.

A few new names emerged in the literature of electron microscopy. One of them was Recknagel. He had published earlier, jointly with Brüche, but in 1935 he published his first major contribution on the relation between electron lens and the electron mirror, jointly with Henneberg.[51] There had been earlier papers on electron mirrors, but I am inclined to select this as one of the most fundamental.

---

50. O. Scherzer, "Über einige Fehler von Elektronenlinsen," *Z. Phys.* 101: 593-603, 1936.

51. W. Henneberg and A. Recknagel, "Zusammenhänge zwischen Elektronenlinse, Elektronenspiegel und Steuerung," *Z. Phys.* 16: 621-623, 1935.

Development in Other Countries

Other indications that the field was maturing were attempts made in three other countries to enter the game. In Britain, Martin, Whelpton, and Parnum, in collaboration with the Metropolitan-Vickers Co., developed a two-stage magnetic microscope.[52] The lenses were entirely inside the vacuum, with external controls. An innovation was the large disk-shaped stage, which allowed a light-optical inspection of the specimen by rotation of the large disk and positioning of the specimen either in the electron beam or in front of the light-optical elements.

In the United States, McMillen and Scott built a simple microscope for biological observation of objects after their calcination.[53] They deliberately destroyed their specimens, hoping to leave a skeleton, which could then be used as a source in an emission microscope.

Last but not least I should like to mention Cotte in Paris, whose first paper on achromatization of electron-optical systems appeared in 1937.[54] Together with Louis Cartan he made the earliest serious attempts in France to do electron-optical research. Cartan was one of the most promising young researchers entering this field; his untimely death during the war of 1939-1945 (he was executed for his part in the French underground resistance against the German occupation) deprived France of one of her best young scientists.

During the year 1937 my first successful bacterial micrographs (Fig. 7) were published.[55] In my first attempts, some of the failures were due to insufficient washing of the specimens and to crystals of inorganic salts masking the essential features. My lack of biological training was only too apparent.

With the appearance of several interesting theoretical papers treating the possibilities of reducing the aberrations of magnetic lenses, I also became interested in lens field distribution.

Gradually the electron microscope became respectable. The growing interest of biologists was noticed by manufacturers of electrical instruments and several companies came to the conclusion that there might be a market for such instruments. I know of four such companies.

---

52. L. C. Martin, R. V. Whelpton, and D. H. Parnum, "A new electron microscope," *J. Sci. Instr.* 14: 14-24, 1937.

53. J. H. McMillen and C. H. Scott, "A magnetic electron microscope of simple design," *Rev. Sci. Instr.* 8: 288-290, 1937.

54. M. Cotte, "Recherches sur l'optique électronique," *Ann. de physique* 10: 333-405, 1938.

55. L. Marton, "La microscopie électronique des objets biologiques: V," *Bull. Acad. Roy. Belg.* 23: 672-675, 1937.

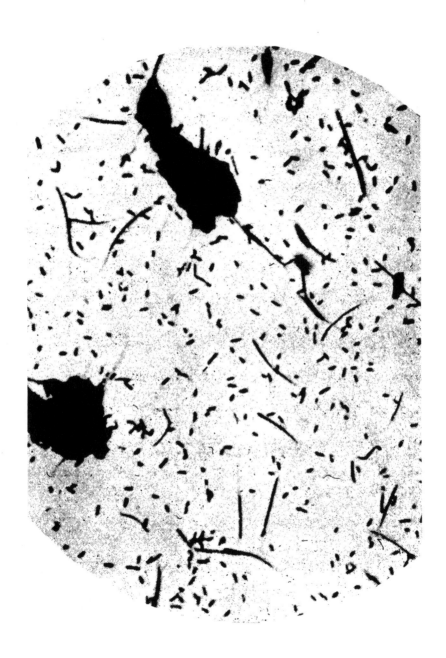

FIG. 7.—First observation of a bacteriological specimen.

Probably the first was Metropolitan-Vickers, which built the Martin, Whelpton, and Parnum instrument as a commercial venture. However, the development was not pushed along for about seven years and they had to start afresh after that interval.

The Siemens effort was more continuous. They hired Ruska and von Borries and put considerable resources at their disposal. They established an effective group of talented young engineers and the results were the production and gradual improvement of the first commercial electron microscope.

Radio Corporation of America also became interested. The laboratory directed by Vladimir Kosma Zworykin had been engaged for several years in electron-optical studies and had achieved considerable success in fields related to television and electron multipliers.

Last, I would like to mention a Belgian optical company, Societé Belge d'Optique in Ghent, whose owner was a distinguished scientist, André Callier. After conversations with him, I started developing plans for a commercial instrument. His untimely death put a stop to that company's efforts.

The general requirements for such an instrument were recognized rather early. A microscope, to be useful and capable of optimum performance, need not necessarily be the technically most advanced model. It must be reasonably rugged, because the user may not have the necessary training for handling a delicate physical apparatus; its operation should be sufficiently "fool-proof." At the same time, it must be sufficiently flexible to give the best possible results in the hands of a skilled operator.

The first Siemens microscope of von Borries and Ruska was a three-lens magnetic device with airlocks for the specimen and for photographic plates (Fig. 8). The very first model still had a cold-cathode gun, but this source of electrons was soon replaced by a thermionic-cathode gun. The gun, the condensor lens, and the object chamber were provided with means for alignment; the objective and projector lenses were fixed in position. All lenses, as mentioned before, had water cooling to compensate for the high power dissipation and to achieve the necessary current constancy. All coils and the gun filament were battery powered. Through careful design of the polepieces, the focal distance of the objective could be reduced to 2.8 mm at 80 keV and of the projector, to 1 mm. The first paper on the Siemens instrument estimated a resolving power of 100 Å.[56]

The first serious analysis of the limitations imposed on the resolving power was published by von Ardenne. In his paper, published early in 1938, he gave numerical estimates for the diffraction defect, the effect of

---

56. B. v. Borries and E. Ruska, "Vorläufige Mitteilung über Fortschritte im Bau und in der Leistung des Übermikroskops," *Wiss. Veröff. Siemens Werken* 17: 99-106, 1938.

FIG. 8.—Earliest commercial electron microscope.

space charge, the aperture defect of the objective, the chromatic defects due to multiple causes, the defects due to external perturbing magnetic fields, and the defects due to volume scattering in the object.[57] For the chromatic defects he considered the following causes: (1) fluctuation of the accelerating voltage, (2) variation of the initial velocities, (3) velocity distribution of the electrons after scattering by the object and its supporting foil, and (4) fluctuations of the currents in the magnetic lenses.

It was a useful compilation, although part of the information was already available elsewhere. Thus, for instance, his considerations for contrast control are a repetition of my paper published two years earlier,[47] without any reference given. In his next paper, von Ardenne described an interesting innovation: the scanning electron microscope.[58] The techniques necessary for its operation were to a considerable extent taken over from television practice. In analyzing the expected performance of the instrument he predicted a resolution of 20 Å at 100 keV. Four years later Zworykin et al., reporting on actual tests, found 500 Å and said that "the limit of performance . . . may be expected to lie in the neighborhood of 100 Å."[59] This is an interesting estimate, because in his first practical results von Ardenne had optimistically already claimed 100 Å, or to be more exact, he claimed that his electron probe had a diameter of 100 Å.

The way was now open for the development of a number of related instruments, such as the shadow microscopes. These are instruments in which an inverted electron microscope makes a highly reduced image of the electron source. The resulting electron image can serve as a virtual source of electrons, and because of its small dimensions, a central projection can produce a highly magnified "shadow" without optical elements, provided the virtual source is very close to the object.

First came the electron shadow microscope of Boersch,[60] then the x-ray shadow microscope of von Ardenne.[61] It is amusing to know that I too had a paper ready and in editorial circulation on this latter subject, but von Ardenne's communication came out first and I withdrew mine. I

---

57. M. v. Ardenne, "Die Grenzen für das Auflösungsvermögen des Elektronenmikroskops," *Z. Phys.* 108: 338-352, 1938.

58. M. v. Ardenne, "Das Elektronen-Rastermikroskop," *Z. Phys.* 109: 553-572, 1938; and *Z. techn. Phys.* 19: 407-416, 1938.

59. V. K. Zworykin, J. Hillier, and R. L. Snyder, "A scanning electron microscope," *ASTM Bull.* 15-23, August 1942.

60. H. Boersch, "Das Schatten-Mikroskop, ein neues Elektronen-Übermikroskop," *Naturwiss.* 27: 418, 1939; and "Das Elektronen-Schattenmikroskop I," *Z. techn. Phys.* 20: 346, 1939.

61. M. v. Ardenne, "Zur Leistungsfähigkeit des Elektronen-Schattenmikroskops und über ein Röntgenstrahlen-Schattenmikroskop," *Naturwiss.* 27: 485-486, 1939.

am mentioning this fact to show that the field now became quite competitive and pretty soon the priority fights started.

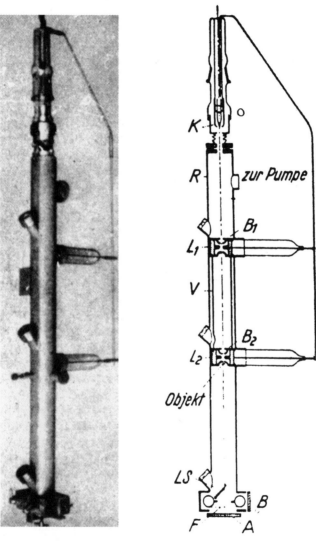

K   Kathode,
L₁ L₂ = Elektrostatische Linsen,

V = Verbindungsrohr
F = Film,

FIG. 9.—First electrostatic microscope (Mahl).

My Work in America

In the fall of 1938 we moved to the United States, where I joined the RCA Laboratories and at the same time took up a lecturership at the University of Pennsylvania. The year 1939 was occupied by the construction of the RCA Type A instrument, which became my No. 4 microscope. Concurrently we made a number of tests with my third microscope, which I had brought over from Brussels.

During that time work also started at the University of Toronto.[62] Prebus and Hillier constructed an instrument with a hot cathode as a source; condensor, objective, and projector lenses; and adjustment provided for a good alignment of the optical path. No airlocks were provided at first; later an airlock was added for the photographic plates. This instrument had surprisingly good resolution.

The same year saw the creation of the first electrostatic transmission microscope (Fig. 9.).[63] Built by Mahl in the AEG laboratories, it was expected to compete with the magnetic instrument. One of the main reasons for its conception may have been the patent situation, since the Siemens group had by that time a firm hold on the magnetic device. This is pure hindsight on my part and should not be interpreted as belittling Mahl's achievements. He used the instrument with great imagination, and before long invented the replica method for bulk specimens.[64] He also pointed out the great advantage of the electrostatic instrument—that no stabilized power supplies are needed for its operation. It is unfortunate that the many other disadvantages (such as easy contamination of the electrodes, with resulting instability of the beam, or inherently greater optical aberrations) outweigh the advantages.

The description of my fourth microscope, the RCA Type A, did not appear until the summer of 1940.[65] It had several features distinguishing it from its predecessors (Fig. 10). At least three of its innovations reappeared years later as "new" inventions. Its gun was probably the first telefocus gun. It had exactly the same spherical configuration as the gun

62. A. Prebus and J. Hillier, "The construction of a magnetic electron microscope of high resolving power," *Canad. J. Res.* (A)17: 49-63, 1939.

63. H. Mahl, "Über das elektrostatische Elektronenmikroskop hoher Auflösung," *Z. techn. Phys.* 20: 316-317, 1939.

64. H. Mahl, "Ein plastisches Abdrucksverfahren zur Übermikroskopischen Untersuchung von Metalloberflächen," *Metallwirtsch., Metallwiss., Metalltechn.* 19: 488-491, 1940.

65. L. Marton, "A new electron microscope," *Phys. Rev.* 58: 57-60, 1940; and L. Marton, M. C. Banca, and J. F. Bender, "A new electron microscope," *RCA Rev.* 5: 232-243, 1940.

FIG. 10.—RCA Type A (foreground), Brussels microscope No. 3 (in back).

described 8 years later by Bruck and Bricka.[66] The objective lens was radically different, as shown by the following description, quoted from my 1940 paper:

> Preliminary experience has shown that it is possible to obtain short focal length with polepieces of considerably greater bore diameter than has been previously accepted. Such a large inner diameter has multiple advantages: the spherical aberration for equal incident beam section is much reduced, the intensity of the image is increased, and the accuracy requirements for machining and aligning are not as high as for smaller bores. The actual polepieces of the objective show an f_{min}/d_p ratio of 0.3 instead of 0.8 which was given in an article by Ruska, for equal velocity and polepiece diameter.[65]

This lens has reappeared in recent publications by Ruska. To quote from a 1964 paper by Ruska:

> More than 20 years ago the physicist Walter Glaser, lately deceased in Vienna, designed such an "optimum electron objective" and calculated its focal length and aberrations. It is a strong magnetic lens: the first half of its field is used as condenser and the second half as part of the objective. In this lens, the specimen must lie in the position at which the axial field strength is a maximum. The spherical aberration of such a field distribution, which decays from a maximum, is about ten times smaller and the resulting resolution limit nearly two times better than in the magnetic objectives used up to now, where the specimen is generally situated in front of the whole field. Though their low aberration has been known for more than 20 years, such "condenser-objective single-field lenses" have not been used in electron microscopes up to now. . . . Such an objective has therefore been built in our institute and was initially tested on an electron-optical bench (Riecke, Stecklein and the author).[67]

It is not my intention to start priority claims, in which electron microscopy is uncomfortably rich. It may be sufficient to mention that a more detailed analysis of this type of operation was the subject of a joint paper[68] with my former student, R. G. E. Hutter (now professor at Brooklyn Polytechnic Institute), in which we investigated in detail the optimum position of an aperture for such a lens. This paper appeared early in 1944.

A third innovation was a new way of combining electron- and light-microscopical observation. It turned out later that the electron microscopist did not need that combination, but later still it became a very useful adjunct of the electron-probe microanalyzer. It consisted of a scaled-up light-microscope objective, provided with an axial bore through which the electron beam could pass.

---

66. H. Bruck and M. Bricka, "Sur un nouveau canon électronique pour tubes à haut tension," *Ann. radioél.* 3: 339-343, 1948.

67. E. Ruska, "Reflections on the past and future development of the electron microscope," in *From Molecule to Cell: Symposium on Electron Microscopy*, Rome: Consiglio Nazionale delle Ricerche, 1964, pp. 77-98.

68. L. Marton and R. G. E. Hutter, "On apertures of transmission type electron microscopes using magnetic lenses," *Phys. Rev.* 65: 161-167, 1944.

Other innovations gained less popularity. The whole optical system, including the coils sealed in copper cans, was mounted inside the outer body of the microscope and was connected by heavy brass cylinders. Thus the optical system acquired a great rigidity and with it a relative freedom from external vibrations. The design of the condensor and projector lenses did not differ greatly from previous designs.

Another advantage of the objective-lens design lay in the introduction of the object, which was placed between two flat apertured brass blades in the region between the polepieces of the objective. This arrangement was partly inspired by two theoretical papers, one by Rebsch and Schneider[69] and one by Glaser.[70]

Another innovation was the power system. Until then all electron microscopes had conventional power supplies for both the high voltage and for the lens currents. At RCA, Arthur Vance was able to develop a new concept by using highly regulated power supplies, with corresponding reduction of the troubles arising from the conventional ones.[71]

During the prewar years there was a time of growing friendship with the USSR. The result was that American industry accepted a certain number of Russian engineers for a training period of several months. In 1940 a young man was assigned to me named V. Goltzov, who spent several months in learning all he could about electron microscopes. He published a resumé in a Russian-language magazine devoted to progress in American technology, but then the war came and he was reassigned to a Russian supply mission taking care of their share of Lend-Lease. As far as I know, that ended his career in electron microscopy; at least, looking for his name in a recent Russian book on this subject, I could not find any mention of him.

Simultaneously with my fourth microscope, von Ardenne built what he called a "universal" electron microscope for bright field, dark field, and stereo observation.[72] Using extremely fine mechanical work, he succeeded in reducing the objective focal length to 1.6 mm and that of the projector to 1 mm. For focusing purposes he used single-crystal ZnS screens, observed through light microscopes, as well as a separate monitoring system in which a sealed-off tube, operated from the same power supplies as the microscope, measured the ratio of high voltage to lens cur-

---

69. R. Rebsch and W. Schneider, "Der Öffnungsfehler schwacher Elektronenlinsen," *Z. Phys.* 107: 138-143, 1937.

70. W. Glaser, "Strenge Berechnung magnetischer Linsen der Feldform $H = H_0 / [1 + (z/a)^2]$," *Z. Phys.* 117: 285-315, 1941.

71. A. W. Vance, "Stable power supplies for electron microscopes," *RCA Rev.* 5: 293-300, 1941.

72. M. v. Ardenne, "Über ein Universal-Elektronenmikroskop für Hellfeld-, Dunkelfeld-, und Stereobild-Betrieb," *Z. Phys.* 115: 339-368, 1940.

rent. He abandoned both in his later work. He paid particular attention to good alignment of the system, which accounts for the high quality of the micrographs achieved with it.

Being a very prolific writer, von Ardenne was the first to bring out an entire book devoted to the electron microscope.[73] The same year, 1940, saw also the publication of an interesting review paper on the subject by von Borries and Ruska.[74] A common feature of these two publications is a thorough quantitative presentation of the optical data and limitations.

James Hillier joined RCA in 1940 and started work on the Type B microscope. This became the instrument first marketed by RCA and a great commercial success. Its good alignment system, combined with Vance's excellent power supplies, made it a very useful instrument.[75]

73. M. v. Ardenne, *Elektronen-Übermikroskopie*, Berlin: J. Springer, 1940.

74. B. v. Borries and E. Ruska, "Mikroskopie hoher Auflösung mit schnellen Elektronen," *Ergebn. ex. Naturw.* 19: 237-322, 190.

75. V. K. Zworykin, J. Hillier, and A. W. Vance, "An electron microscope for practical laboratory service," *Elec. Eng.* 60: 157-161, 1941.

Single Instead of Multiple Scattering

At about the same time I had the good fortune of becoming acquainted with Leonard Schiff. With his outstanding knowledge of quantum mechanics we jointly produced the first analysis of the scattering phenomena in the electron microscope, using a modern viewpoint.[76] This work started a string of papers in the following years, all improving on our original one.

I have mentioned priority squabbles. The war years of 1939-1945 produced one of the worst priority fights that I have ever seen, in Germany. The protagonists were von Borries and Ruska on the one hand and Brüche on the other. Claims and counterclaims were fired by the opposing camps in a number of papers. Part of the quarrel now appears to have been based on semantic misunderstandings. It took almost five years before the two opposing groups signed a truce. I call it a truce, although with AEG's subsequent complete withdrawal from the race it may be qualified as a victory for Siemens. The Siemens laboratories survived the war without serious losses, whereas the AEG laboratories were severely affected, which might have been one of the reasons for transferring these activities after the war to the Zeiss group that fled from Jena in East Germany to Oberkochen in West Germany at about that time.

But let us come back to the early war days, which produced two very important results, both from the AEG laboratories. One was Recknagel's theory of the emission microscope.[77] He succeeded in demonstrating that the resolving power depends essentially on the energy distribution of the electrons and on the accelerating field at the cathode when the electrons leave it. I mentioned earlier some of the speculations about the effect of contact potentials. Mecklenburg succeeded in the experimental demonstration that Recknagel's theory was right and that the contact potentials had less effect than originally feared.[78]

The second important event was the discovery of Fresnel diffraction, which was apparently observed simultaneously at the AEG and RCA laboratories. Boersch in Berlin[79] and Hillier in Camden, N.J.,[80] both published notes in 1940 on the fringes in electron-microscope images, which

---

76. L. Marton and L. I. Schiff, "Determination of object thickness in electron microscopy," *J. Appl. Phys.* 12: 759-765, 1941.

77. A. Recknagel, "Theorie des elektrischen Elektronenmikroskops für Selbststrahler," *Z. Phys.* 117: 689-708, 1941.

78. W. Mecklenburg, "Über das elektrostatische Emissions-Übermikroskop," *Z. Phys.* 120: 21-30, 1942.

79. H. Boersch, "Fresnelsche Beugung im Elektronenmikroskop," *Naturwiss.* 28: 709, 1940; and "Randbeugung der Elektronen," *Phys. Z.* 44: 32-38, 202-211, 1943.

until then were attributed to other causes, such as chromatic aberration. The recognition that they are caused by interference phenomena was greatly helped by the then improved resolution of the new instruments. In the following years ample proof was found in both laboratories for the Fresnel fringe interpretation.

80. J. Hillier, "Fresnel diffraction of electrons as a contour phenomenon in electron supermicroscope image," *Phys. Rev.* 58: 842, 1940.

The Stanford Microscope

The war years slowed down the development somewhat but did not stop it altogether. Shortly before the war I moved to Stanford University, where a new Division of Electron Optics was set up with the help of the Rockefeller Foundation. The first task was to design and construct a new electron microscope.[81] Several features distinguished this instrument from its predecessors. Although some attempts have been made in the past to subdivide either the objective or the projector lens, neither of them allowed a sufficient spacing for use of the components in more than an auxiliary way. The Stanford microscope (Fig. 11) deliberately used a three-stage magnification of the image, a design that has been very popular ever since. Practically all the microscopes put on the market in the postwar years are provided with intermediate lenses. Another feature was the use of a double condensor lens. The stage motion was controlled by a hydraulic system, a method that may be now, a quarter of a century later, revived in the proposed high-voltage instrument at the Argonne National Laboratory. Another feature was adopted sooner. I refer to the displacement of the polepieces for alignment purposes, rather than displacement of the whole lens assembly. This feature was later incorporated in the high-voltage microscope of Gaston Dupouy in Toulouse.[82] The Stanford instrument could not have been built without the enthusiastic help of R. G. E. Hutter and B. F. Bubb.

An interesting consequence of the work at Stanford was the acquisition of a guest worker from India. During a visit of Prof. M. N. Saha (1893-1956) in 1943 we agreed to have an Indian physicist accepted on a fellowship basis. Dr. (now Professor) N. N. Das Gupta came for a year, with our help built a horizontal electron microscope, and took it home to Calcutta. During a recent visit I was greatly pleased to see that the instrument is still in operating condition after 22 years.[83]

A little earlier I mentioned Rudolf Hutter, who was my collaborator (and graduate student) during the Stanford years. His stay was quite productive; besides his thesis work on an electrostatic lens, he produced several other very interesting papers on electron-optical subjects. Another graduate student was K. Bol, with whom we investigated the possibility

81. L. Marton, "A new world beyond," *Stanford Alumni Rev.* 44(No. 8): 7-9, 1943; "A 100 kV electron microscope," *Science* 100: 318-320, 1944; and "A 100 kV electron microscope," *J. Appl. Phys.* 16: 131-138, 1945.

82. G. Dupouy, "Microscope électronique à très haute tension," *Ann. de Physique* 8: 251-260, 1963.

83. N. N. Das Gupta, M. L. De, D. L. Battacharya, and A. K. Chaudhury, "A new horizontal electron microscope," *Indian J. Phys.* 22: 497-513, 1948.

FIG. 11.—Stanford microscope.

of reduction of spherical aberration in magnetic lenses. We succeeded in demonstrating some reduction of the aberration, but not to a degree required in electron microscope applications.

One more attempt at the reduction (or elimination?) of spherical aberration had its origin during those years. In 1942 I speculated on the possibility of using diamagnetic components in a rotationally symmetric magnetic field for correction of the aberration. A patent was granted in 1945 on this subject, but the very limited range of diamagnetic susceptibilities of common substances made any attempt at the practical realization illusory. The recent availability of hard superconductors (perfect diamagnetic bodies) makes the whole scheme much more reasonable. I am convinced that within the next few years we shall be able to produce magnetic lenses with a very high degree of correction.

During the war years an interesting attempt was made at the General Electric Co. to produce and market a simple electrostatic microscope. C. H. Bachman and Simon Ramo made a thorough study of such a system[84] and produced an extremely simple instrument, with a resolution limited to 200 Å. I do not know what prevented the company from marketing it.

During the Stanford years, we became very good friends with the colloid chemist Sydney Ross (nor professor at Rensselaer Polytechnic Institute). We had very good fun in concocting a Sigma Xi lecture entitled "Alice in Electronland."[85] The fun did not altogether stop with the preparation and presentation of the lecture. Several years later I was still getting inquiries asking whether that paper had been a hoax or not!

The need was felt for a bibliography of electron microscopy. Mrs. Marton undertook the task in collaboration with Samuel Sass (librarian of the Physics Department at the University of Michigan and later librarian of the General Electric Co. in Pittsfield). It came out in three installments[86] and became a useful tool for the growing number of electron microscopists.

Of the postwar proliferation of instruments, methods, and accessories I should like to mention a few in my closing remarks. The Philips microscope, designed by Le Poole, was the first commercial microscope incorporating the three-stage magnification system. A very clever innovation was the utilization of the added optical combination for "selective area diffraction." This application of Boersch's principle became a permanent

---

84. C. H. Bachman and S. Ramo, "Electrostatic electron microscopy," *J. Appl. Phys.* 14: 8-18, 1943.

85. L. Marton, "Alice in Electronland," *Am. Scientist* 31: 247-254, 1943.

86. C. Marton and S. Sass, "A bibliography of electron microscopy," I: *J. Appl. Phys.* 14: 422-531, 1943; II: 15: 575-579, 1944; III: 16: 373-378, 1945.

feature of all modern instruments.[87]

An interesting attempt was made to facilitate focusing by means of a "wobbler." For those who may have forgotten this ingenious device, it consisted in a superposition of an ac deflecting field, so arranged that the two components of the image fused together only for the right focusing condition. Its use was limited to the Philips instrument.

I should like to pick out another item as a significant development of the immediate postwar years. I refer to the recognition by Hillier and Ramberg of the influence of astigmatism on the resolution and their method for correcting it.[88] This again, with modifications, became a standard feature in all microscopes.

Last but not least, I should like to include two important contributions from people I have mentioned repeatedly. One is a 1947 paper by Scherzer on the correction of spherical and chromatic aberrations. His theorem on the possible use of two-dimensional correcting elements may be considered as the basis of all modern development of quadrupole and octupole optics.[89]

Soon thereafter came a short notice by Gabor on microscopy by reconstructed wave fronts.[90] Its application in electron microscopy is still not accomplished, but I am among those who think that this is merely a question of time. The wide development of optical holography since lasers became available indicates how fruitful the concept is.

The French effort got under way in a three-pronged approach. Two of the efforts were linked to commercial ventures, the third was purely academic. Dupouy helped the optical company OPL to develop a magnetic instrument.[91] Pierre Grivet jointed the radio company CSF and designed an electrostatic microscope.[92] The third venture started at the École Polytechnique, where Claude Magnan and Pierre Chanson initiated the construction of a proton microscope.[93]

In Great Britain, Metropolitan-Vickers resumed work on the electron

---

87. J. B. Le Poole, "A new electron microscope with continuously variable magnification," *Philips Tech. Rev.* 9: 33-45, 1947.

88. J. Hillier and E. G. Ramberg, "The magnetic electron microscope objective," *J. Appl. Phys.* 18: 48-71, 1947.

89. O. Scherzer, "Sphärische und chromatische Korrektur von Elektronenlinsen," *Optik* 2: 114-132, 1947.

90. D. Gabor, "A new microscopic principle," *Nature* 161: 777, 1948; "Microscopy by reconstructed wave-fronts," *Proc. Roy. Soc.* (A)197: 454-487, 1951; and "Microscopy by reconstructed wave fronts: II," *Proc. Phys. Soc.* (B)64: 449-469, 1951.

91. G. Dupouy, "Microscope électronique magnétique à grand pouvoir de résolution," *J. Phys. et Rad.* (8)7: 320-329, 1946.

92. P. Grivet, "Le microscope électrostatique C.S.F.," *Le vide* 1: 29-47, 1946; and P. Grivet and H. Bruck, "Le microscope électronique électrostatique," *Ann. Radioél.* 1: 293-310, 1946.

microscope under the leadership of Michael Haine.[94] He succeeded in gathering an able group. I should like to mention particularly G. Liebman, T. Mulvey, P. A. Einstein, D. E. Bradley, and G. D. Archard.

Another important center grew up at Cambridge University under the leadership of V. E. Cosslett, author of an important textbook.[95] Many excellent workers in electron optics, such as W. Nixon and others, were trained by Prof. Cosslett and his collaborators and he contributed markedly to the growth of the field. One outstanding man, P. A. Sturrock (who in turn authored an important monograph[96]), worked for a time under my direction and also under Prof. Grivet.

Japanese work started during the war. T. Hibi was probably the first to build an instrument, with B. Tadano close second.

Efforts in the Soviet Union apparently followed the Western development. Design of recent Soviet instruments is very similar to the Western ones. Surprisingly the instrument omits the use of airlocks, both for the specimens and for the photographic material.

The few years preceding the war were quite fertile in that several other important phenomena were recognized. I mentioned diffraction: Boersch[49] was the first to point out the important correlation between diffraction and image formation, which later led to the development of the "selected area diffraction: observation."[97] Shortly thereafter Boersch recognized that, with sufficient resolution of the instrument and a small enough numerical aperture, Fresnel fringes can be observed in the out-of-focus image, and important observation that did not escape the contemporary workers.[79]

Previous to the recognition of the Fresnel fringes, there was quite a bit of speculation in the literature about the origin of some of the strange edge effects observed in electron microscope images. Hillier[98] assumed them to be due to chromatic defects, whereas von Borries and Ruska[99]

---

93. C. Magnan, P. Chanson, and A. Ertaud, "Sur un projet de microscope protonique," *Compt. rend.* 220: 770-772, 1945; and "Optique corpusculaire," *ibid.* 238: 1701 and 1797, 1954.

94. M. Haine, "The design and construction of a new electron microscope," *J. IEE* 94: 447-462, 1947.

95. V. E. Cosslett, *Introduction to Electron Optics*, Oxford: Clarendon Press, 1946.

96. P. A. Sturrock, *Static and Dynamic Electron Optics*, London: Cambridge University Press, 1955.

97. A. C. van Dorsten, J. W. Oosterkamp, and J. B. Le Poole, "An experimental electron microscope for 400 kilovolts," *Philips Tech. Rev.* 9: 193-201, 1947.

98. J. Hillier, "The effect of chromatic error on electron microscope images," *Canad. J. Res.* 17(A): 64-69, 1939.

99. B. v. Borries and E. Ruska, "Eigenschaften der Übermikroskopischen Abbildung," *Naturwiss.* 27: 281-287, 1939.

attempted an explanation based on spherical aberration.

In 1940 both Boersch[79] and Hillier[80] suspected (without adequate proof) that the effects might be due to Fresnel diffraction, but definite proof was given only by Boersch's[79] paper published in 1943. Shortly thereafter, Hillier and Ramberg[88] published some computations of Fresnel patterns and compared them with observed microphotometer traces. The necessary introduction of phase delays led to considerations of phase microscopy and became the foundation for all the future important work of Gabor.[90]

Immediately after the war a very good book came out on electron optics and electron microscopy. I refer to the book by Zworykin and collaborators,[100] which remained for many years the standard text on the subject and whose multiple authorship earned it a sobriquet taken from the preamble of the U.S. Constitution: in American electronics laboratories, the book was always referred to as "We, the People."

The title of this survey implies that I should limit myself to the early history of the microscope. The cut-off date of the immediate postwar years is arbitrary and that is why it may be legitimate to say a few words about the postwar years. I mentioned a few of these later developments in passing and would now like to amplify these remarks slightly. Two good centers in Britain contributed quite a bit; their accomplishments, together with those of Gabor, were related earlier. In France, both Dupouy and Grivet created fertile schools, with the electron-probe microanalyzer of R. Castaing becoming a feature. In Germany, E. Ruska was very successful in creating a large and devoted group; H. Boersch joined the Technical University of Berlin where, together with his students, he remains one of the most prolific contributors to electron physics. This report would not be complete without mentioning the important schools initiated by H. Raether in Hamburg and G. Möllenstedt in Tübingen.

In the USSR, some of the important names are: V. M. Kel'man, M. I. Korsunskyi, O. I. Seman, and S. Y. Yavor.

In the U.S., instrument development remained linked with work at RCA, with the late R. G. Picard and later J. H. Reisner and co-workers leading the field.

Very extensive instrument evolution took place during the after-war years in Japan. Some of the outstanding names are H. Hashimoto, K. Kanaya, K. Ito, Y. Sakaki, R. Uyeda, and H. Watanabe.

---

100. V. K. Zworykin, G. A. Morton, E. G. Ramberg, J. Hillier, and A. W. Vance, *Electron Optics and the Electron Microscope*, New York: Wiley, 1945.

Envoi

This enumeration is far from complete. The intention was rather to call attention to the very extensive literature that has grown up during the past twenty years. (Several surveys are mentioned in a short Bibliographical Note that follows the Acknowledgments.)

Nor does the above account claim to be complete or "objective." First, it is an intention of this series of monographs to record a personal account of a participant in some events in the development of modern technology, and that I have done. Besides, I do not believe that this so-called objectivity really exists. Whichever way we look at any happening, we introduce a certain bias and it is doubtless better to acknowledge our bias than to hide it. It gives the reader a better chance to apply corrections. I read recently a paper by Stanley L. Jaki, a historian of science. He writes that "some scientists can be shockingly careless when it comes to the presentation of a detail of scientific history, a matter closely related to scientific research."[101] I believe this "carelessness" is partly due to our limited ability of reading and retaining. A few months ago I was offered a rather startling demonstration. I was listening to a paper presented by a very able young physicist, and afterwards in a private conversation, I asked him in what detail or manner his work differed from something I happened to publish 25 years earlier. He was not aware of the earlier work, but was very interested in it. Our conversation was overheard by an even younger man, who interrupted us: "Who can read all that old stuff?"

Yes, that may be the question! Who can read all the old stuff? Yet I do not know how progress can be conceived without at least a limited knowledge of past performance and past achievements. I think that the remark of my young friend should not be taken very seriously. The existence of this lecture series, to which I have the honor to contribute, is an eloquent proof of the interest in the past, an interest that looks forward and wishes to base tomorrow's achievements on a solid foundation.

It is not my intention to moralize, but I think we can draw a conclusion from our contemplation of the subject presented here. As in any other field, progress does not necessarily go by leaps and bounds. Every step has its predecessor and more often than not, the more important steps are reached simultaneously by two or more persons. We often say a new idea is or was "in the air," meaning usually that on the basis of previous knowledge, it is surprising that someone else did not think of what one may have thought of. Science has always been a continuous process

---

101. S. L. Jaki, "Olbers', Halley's, or whose paradox?" *Am. J. Phys.* 35: 200-210, 1967.

and this story may be a good illustration of how difficult it sometimes is to make a definite statement.

What I have said amounts to this. Legally, Rüdenberg is the inventor of the electron microscope because the patent documents on file say so. Practically, the electron microscope is the creation of a number of people and a few names stand out. If I insisted in my presentation on giving you more than a few names, it is because I feel strongly that the minor contributions should not be forgotten. The day may come when someone will write a new kind of history of science, where instead of extolling the contributions of the giants, he will show us in detail the contributions of the forgotten.

Since this material was first presented in Berkeley, I cannot end my contribution more appropriately than with a quotation from one of Berkeley's great men, now departed from the scene, J. Robert Oppenheimer (1904-1967). In an address before the Tenth Anniversary Conference of the Congress for Cultural Freedom in Berlin in June 1960, he touched on tradition, this time introducing one of his favorite themes—the need for common understanding. He said:

I have been much concerned that in this world we have so largely lost the ability to talk with one another. In the great succession of deep discoveries, we have become removed from one another in tradition, and in a certain measure even in language. We have had neither the time nor the skill nor the dedication to tell one another what we have learned, nor to listen nor to hear, nor to welcome its enrichment of the common culture and the common understanding. Thus the public sector of our lives, what we have and hold in common, has suffered, as have the illumination of the arts, the deepening of justice and virtue, the ennobling of power and of our common discourse. We are less men for this. Our specialized traditions flourish; our private beauties thrive; but in these high undertakings where man derives strength and insight from the public excellence, we have been impoverished. We hunger for nobility: the rare words and acts that harmonize simplicity and truth. In this I see some connection with the great unresolved public problems: survival, liberty, fraternity.[102]

---

102. J. R. Oppenheimer, "A time of sorrow and renewal," *Encounter* 16: 70-72, 1961.

Acknowledgments

I am greatly indebted to my wife, Claire Marton, for her help in preparing this survey. Her recollection of past events, often better than mine, contributed greatly to making this presentation better balanced.

I also wish to express thanks to Dr. Bernard Finn, Curator for Electricity at the Museum of History and Technology,* Smithsonian Institution, Washington, D.C. As most of my early records, negatives, and other materials are permanently deposited at the Museum, I acknowledge the help given by the Smithsonian Institution, especially with regard to some of the illustrations.

Bibliographical Note

In writing these reminiscences I used my personal notes, the literature cited, and several papers published on the history of the electron microscope. The most important historical papers perused are listed below alphabetically by author.

B. v. Borries and E. Ruska, "Mikroskopie hoher Auflösung mit schnellen Elektronen," *Ergebn. ex. Naturw.* 19: 237-322, 1940; and "Neue Beiträge zur Entwicklungsgeschichte der Elektronenmikroskopie und der Übermikroskopie," *Phys. Z.* 45: 314-326, 1944.

E. Brüche, "Zum Entstehen des Elektronenmikroskops," *Phys. Z.* 44: 176-180, 1943; and "Forschungsreise durch das Gebiet der Elektronenoptik," *Optik* 24: 290-295, 1966-67.

M. M. Freundlich, "Origin of the electron microscope," *Science* 142: 185-188, 1963.

D. Gabor, "Die Entwicklungsgeschichte des Elektronenmikroskops," *ETZ* 78: 522-530, 1957.

K. Kupfmüller, "Zur Geschichte des Elektronenmikroskops," *Phys. Z.* 45: 47-51, 1944.

A. Matthias, "Bemerkung zur Entstehung des Elektronenmikroskops," *Phys. Z.* 43: 129-131, 1942.

T. Mulvey, "Origins and historical development of the electron microscope," *Brit. J. Appl. Phys.* 13: 197-207, 1962; and "The history of the electron microscope," *Proc. Roy. Micr. Soc.* 2(Pt. 1): 201-227, 1967.

R. Rüdenberg, "The early history of the electron microscope," *J. Appl. Phys.* 14: 434-436, 1943.

E. Ruska, "25 Jahre Elektronenmikroskopie," *ETZ* 78: 531-543, 1957.

---

*Now the National Museum of American History.

Postscript (1994)

As noted, L. L. Marton's account is based on a lecture he gave at Berkeley in 1967. It was published by San Francisco Press in its *History of Technology Monographs* series in 1968; the present version is a second edition. The book also formed the basis for the undersigned's chapter on Marton (with a complete list of his publications) in *The Beginnings of Electron Microscopy* (1985), edited by P. W. Hawkes as Supplement 16 to his Academic Press series *Advances in Electronics and Electron Physics.* (Marton had been its editor for the first 55 volumes, jointly with his wife Claire after 1958.)

Marton had become Principal Physicist at the National Bureau of Standards (now the National Institute of Standards and Technology) in Washington in 1946 and in 1948 was appointed chief of its newly formed Electron Physics Section, a post he held until 1964, when he was named head of the NBS Office of International Relations. He retired in 1970, only to become an unsalaried research associate at the Smithsonian Institution, where he was able to indulge his life-long interest in the history and philosophy of science and technology. He died in Washington on 22 January 1979, of a heart attack, aged 77.

His good friend and fellow countryman Dennis Gabor (1900-1979), whose Preface to the present volume has been carried over from the first edition, had an even more distinguished career. After a score of years as an electrical engineer in German and British industry, he joined the faculty of Imperial College in London in 1949 (and was named professor of electron physics in 1958), where he made important contributions to electron optics, communication theory, and plasma physics. In 1947 he had conceived the idea of holography, which remained of purely theoretical interest until the development of lasers in the early 1960s. For this contribution Gabor received the 1971 Nobel Prize for physics.

It is generally acknowledged that Marton was the first to use an electron microscope (moreover, one of his own design) to examine biological specimens. In addition, with an insistence uncharacteristic of this most ethical and modest of scientists, Marton claimed another priority for himself: the discovery that the action of the electron microscope did not depend on absorption (as in the optical microscope), but on scattering (pp. 16-18). This claim was noted in the aforementioned tribute to Marton in *The Beginnings of Electron Microscopy,* but a draft of it shown to Ernst Ruska elicited a note pointing out that his old friend Marton, who had based his claim on the exact wording of the sentence quoted on p. 18 of the present volume, had evidently omitted to take into account the sentence immediately preceding the cited one—a sentence in which scattering angles *(Streuwinkel)* were mentioned in defining *Diffusion:*

> On the other hand, in objects that are very thin in relation to beam potential the reason the image intensity varies considerably despite the transmitted beam-current intensity being kept constant is that individual points of the object give rise to rays of nonuniform [optical] aperture (intensity distribution to scattering angles) corresponding to their mass density, so that if only the lens apertures are sufficiently small, currents of differing intensities appear to originate from various points of the object and show up in the image.

Even so, Ruska did acknowledge in his note that immediately following his own brief reference to these matters, it was Marton who pursued them more fully.

In 1957 Gabor had noted how strange it was that the epoch-making invention of the electron microscope had not been marked by the award of a Nobel Prize, especially since all the major contributors except Bodo von Borries (1905-1956) were still alive.[8] (Gabor could not have anticipated that he would receive one himself 14 years later for an altogether different achievement.) Borries and his brother-in-law Ruska were the most obvious candidates, even though the first patent had been taken out in 1931 by R. Rüdenberg.[103] The omission was not repaired until 1986, when the 79-year-old Ernst Ruska unexpectedly received the Nobel Prize for physics. A few years earlier, in a conversation with the undersigned, he had said wistfully that he certainly did not expect one after all that had passed in the more than half a century since he had worked under Max Knoll at the Technische Hochschule in Berlin. Ernst Ruska died in 1988, at the age of 81.

Charles Süsskind

---

103. In 1979 Ruska had published a 136-page monograph, "Die frühe Entwicklung der Elektronenlinsen und der Elektronenmikroskopie," which appeared as No. 12 of *Acta Historica Leopoldiana* of the Deutsche Akademie der Naturforscher Leopoldiana in Halle (in the former German Democratic Republic). It has been translated into English by T. Mulvey, *The Early Development of Electron Lenses and Electron Microscopy*, Stuttgart: Hirzel, 1980. One part, a correspondence with Max Steenback at the University of Jena regarding Rüdenberg's claims to priority of invention of the electron microscope, has been reproduced, in German and English, on pp. 602-608 of the above-mentioned Supplement 16 edited by P. W. Hawkes, *The Beginnings of Electron Microscopy* (New York: Academic Press, 1985).

Name Index

A

Abbe, E. vii, 12
Archard, G. D. 45
Ardenne, M. v. 30, 32, 37, 38
Auger, P. 25

B

Bachman, C. H. 43
Banca, M. C. 34
Bender, J. F. 34
Berthod, A. R. 3
Birkeland, K. O. B. 2
Boersch, H. 26, 32, 39, 43, 46
Bol, K. 41
Borries, B. v. 16, 19, 24, 30, 38,
    39, 45, 49, 51
Bothe, W. 18
Bradley, D. E. 45
Bricka, M. 36
Brien, P. 16
Brüche, E. 4, 7, 12, 13, 19, 21,
    24, 27, 39, 49
Bruck, H. 36
Bubb, B. F. 41
Burgers, W. G. 24
Busch, H. W. H. 3, 4, 13

C

Calbick, C. J. 7
Callier, A. 30
Cartan, L. 28
Castaing, R. 46
Chanson, P. 44
Chaudhury, A. K. 41
Cosslett, V. E. 45
Cotte, M. 28
Crookes, W. 3

D

Darboux, J. G. 2
Das Gupta, N. N. 41
Davisson, C. J. 7

De, M. L. 41
Dorsten, A. C. v. 45
D'Ortous de Mairan, J. J. 3
Driest, E. 26
Dupouy, G. 41, 44, 46

E

Einstein, P. A. 45
Ertaud, A. 45

F

Fermat, S. de 13
Finn, B. 49
Franklin, B. 2
Fresnel, A. J. 39, 40
Freundlich, M. M. 49

G

Gabor, D. 3, 4, 46, 49–51
Gauss, K. F. 2
Germer, L. H. 7, 8
Glaser, W. 13, 15, 36, 37
Grivet, P. 44-46

H

Haine, M. 45
Halley, E. 2
Hamilton, W. R. 4, 13
Hashimoto, H. 46
Hauksbee, F. 3
Hawkes, P. W. 50
Henneberg, W. 27
Henriot, E. 9, 22, 24
Henry, J. 2
Hérelle, F. d.' vii
Herzen, S. 23
Herzog, R. 24
Hibi, T. 45
Hillier, J. 32, 34, 38, 39, 44–46
Homès, M. 16
Houtermans, F. G. 12
Hutter, R. G. E. 36, 41